THE POETRY OF EINSTEINIUM

The Poetry of Einsteinium

Walter the Educator

Silent King Books

SILENT KING BOOKS

SKB

Copyright © 2024 by Walter the Educator

All rights reserved. No part of this book may be reproduced in any manner whatsoever without written permission except in the case of brief quotations embodied in critical articles and reviews.

First Printing, 2024

Disclaimer
This book is a literary work; poems are not about specific persons, locations, situations, and/or circumstances unless mentioned in a historical context. This book is for entertainment and informational purposes only. The author and publisher offer this information without warranties expressed or implied. No matter the grounds, neither the author nor the publisher will be accountable for any losses, injuries, or other damages caused by the reader's use of this book. The use of this book acknowledges an understanding and acceptance of this disclaimer.

"Earning a degree in chemistry changed my life!"
- Walter the Educator

dedicated to all the chemistry lovers, like myself, across the world

EINSTEINIUM

Einsteinium, a marvel, a guide.

EINSTEINIUM

Born in the heart of a nuclear blast,

EINSTEINIUM

Its story unfolds, an enigma amassed.

EINSTEINIUM

With atomic number ninety-nine,

EINSTEINIUM

In the periodic table, it starts to shine.

EINSTEINIUM

Named for a genius, a mind so bright,

EINSTEINIUM

Einstein's legacy, a beacon of light.

EINSTEINIUM

A fleeting creation, elusive, rare,

EINSTEINIUM

In nature's dance, it's exceptionally spare.

EINSTEINIUM

Formed in reactors, where science prevails,

EINSTEINIUM

Unraveling mysteries, it never fails.

EINSTEINIUM

Its properties masked, a mystery deep,

EINSTEINIUM

Invisible to eyes, it silently sleeps.

EINSTEINIUM

A metal of wonder, a substance unique,

EINSTEINIUM

In laboratories, its secrets we seek.

EINSTEINIUM

Einsteinium whispers tales untold,

EINSTEINIUM

Of atomic structure, mysteries unfold.

EINSTEINIUM

Its nucleus teeming with protons and neutrons,

EINSTEINIUM

A dance of particles, where chaos adjourns.

EINSTEINIUM

In the depths of the cosmos, it may reside,

EINSTEINIUM

Amongst celestial wonders, it may glide.

EINSTEINIUM

A cosmic wanderer, in the vast unknown,

EINSTEINIUM

Einsteinium's journey, forever grown.

EINSTEINIUM

Unlocking the secrets of its atomic shell,

EINSTEINIUM

Scientists delve, where mysteries dwell.

EINSTEINIUM

With each discovery, a step towards the light,

EINSTEINIUM

Einsteinium's essence, burning bright.

EINSTEINIUM

In the hands of scientists, a tool of might,

EINSTEINIUM

Unraveling nature's codes, shining bright.

EINSTEINIUM

From medical wonders to energy's embrace,

EINSTEINIUM

Einsteinium's potential, a boundless space.

EINSTEINIUM

Yet caution prevails, for its power immense,

EINSTEINIUM

Handle with care, in its presence, intense.

EINSTEINIUM

For with great knowledge comes great responsibility,

EINSTEINIUM

Einsteinium's legacy, a testament to humility.

EINSTEINIUM

So let us marvel at this element divine,

EINSTEINIUM

Einsteinium's legacy, a treasure to find.

EINSTEINIUM

In the tapestry of the universe, it weaves,

EINSTEINIUM

A testament to human curiosity, it believes.

EINSTEINIUM

From the depths of the atom to the stars above,

EINSTEINIUM

Einsteinium's story, an ode to love.

EINSTEINIUM

For in the quest for knowledge, we find our grace,

EINSTEINIUM

Einsteinium's embrace, a celestial embrace.

EINSTEINIUM

ABOUT THE CREATOR

Walter the Educator is one of the pseudonyms for Walter Anderson. Formally educated in Chemistry, Business, and Education, he is an educator, an author, a diverse entrepreneur, and he is the son of a disabled war veteran. "Walter the Educator" shares his time between educating and creating. He holds interests and owns several creative projects that entertain, enlighten, enhance, and educate, hoping to inspire and motivate you.

> Follow, find new works, and stay up to date
> with Walter the Educator™
> at WaltertheEducator.com

www.ingramcontent.com/pod-product-compliance
Lightning Source LLC
LaVergne TN
LVHW020134080526
838201LV00119B/3883